D0518308

AMERICAN ROADS

AMERICA

N ROADS

Photographs by WINSTON SWIFT BOYER

Introduction by William Least Heat-Moon

Bulfinch Press · Little, Brown and Company · BOSTON · TORONTO · LONDON

I WOULD LIKE TO THANK JOSEPHINE BOYER, TRACY CHESEBROUGH, KATHLEEN MANSON, ERIC HOLDSWORTH, STEVEN HYDE, RICHARD AND AMORY MILLARD, RAY ROBERTS, "ROUTE 66," AND ALL THOSE WHO MADE THIS BOOK POSSIBLE.

—W. S. B.

FIRST EDITION

ISBN 0-8212-1708-9

LIBRARY OF CONGRESS CATALOG CARD NUMBER 88-83461

BULFINCH PRESS IS AN IMPRINT AND A TRADEMARK OF LITTLE, BROWN AND COMPANY (INC.)

PUBLISHED SIMULTANEOUSLY IN CANADA BY LITTLE, BROWN & COMPANY (CANADA) LIMITED

PRINTED IN JAPAN

To my grandmother, Mrs. Lila Leonard Swift

The Yellow Line, Highway 130, North Carolina, 1987

INTRODUCTION *by William Least Heat-Moon*

I live in the woods in the middle of Missouri, and the road that edges our place rises and falls and turns, and it's bordered with sugar maple and red oak, sassafras and hickory. A century and more ago, this lane was a plank road – oak logs squared with an adz on three sides and laid transversely across two lines of parallel timbers. Over some ten miles, the oaken road linked the old university town with the Missouri River. The port was Providence, today a ghost town of a few broken fishing shacks and some well-laid limestone walls, and the plank road, torn up by heavy whiskey wagons making their sloshing way down to the steamboat wharf, is gravel.

I've chosen to live here in the river-hollow country because I like the circuitous course the old road takes among the slopes as it moves almost reluctantly away from the river and toward town. A road, you see, after my own heart. I like the ghost presence of the whiskey wagons knocking over the wooden planks, and I like the legend that Abraham Lincoln had walked the road from wharf to town so he could spark Mary Todd, who visited relatives here one summer. The history and beauty of this old lane draw me in, and my sharing them somehow makes me belong.

But there's more to it than mere nostalgia. This road from another time has an ease about it, a gracefulness, a quiet, a way of fitting the terrain that reveals how it too belongs to these ice-age hills. Coming home along it, I'm put into that ease and grace, and I'm quieted by its presence, its shape, its fusion with the land. It smooths my edges as it awakens me and brings me *into* things. Take that road from me and you take some of what I now am and may become.

When I leave home and head toward Kansas City, where I was born and where most of my people live, I go a mile up the limestone lane, then onto an unstriped asphalt road, onto striped asphalt, onto two-lane concrete, and finally onto a divided four-lane. There's 135 miles of that. I-70 has a little history now also: it's the first interstate to be completed across a state, and it often follows old U.S. 40, which traced in here the first leg of the Oregon and Santa Fe trails. From Columbia, Missouri, to Kansas City, I-70 was once a pleasing highway that opened into the countryside and let the traveler out as if he were being set free.

On the nine hundred and some trips (that's four months of my life) I've made over it in the last third of a century, I've watched that highway close me in and shut me off, bind me in visually and spiritually, its closures increasing at an *increasing* rate with changes visible these days not just from year to year but from monthly junket to junket. Today, I-70 is a tube of the hideous, of vinyl-food franchises, overpriced cut-rate motels, metal self-storage sheds, and miles of monstrous billboards that even at midnight show the traveler twenty-foot-long french fries and twelve-foot-high packs of cigarettes illuminated by electricity from a sulfurous coal furnace somewhere or a dam across the once navigable Osage River. On some of these drives, now so long, I think about the most successful highway advertising ever: eighteen-by-forty-*inch* boards, six little linked signs that won a driver's attention not with giantism and floodlights but with language and wit and sometimes corn to laugh you awake:

A MAN
A MISS
A CAR – A CURVE
HE KISSED THE MISS
AND MISSED THE CURVE
BURMA-SHAVE

These days when I arrive at my parents' home, my eyes and mind and soul abused by the interstate hours, I'm depressed about American rapaciousness and not at all ready for a family visit.

"Don't look," someone advised me. Is that the secret? Psychic survival is anesthesia? Is that what the road despoilers are after, a numbed nation riding down its corridors of putrescence?

Yet, there is another America, endangered perhaps, but, at least for a while, one still there for the road traveler. It's the one Winston Swift Boyer shows in these sixty-four photographs, and it's a place that gives reason for hope and, even more, gives cause for popular rebellion. It's a countryside worth fighting for. These photographs are a manifesto, a call to resist the degradation of the billboard boys who peddle their stuff in violation of federal law, a call to thwart the realty hustlers. Pretty pictures of a country can mislead and lull or they can alert and awaken.

Boyer shows me an America I want to live in and belong to, and his work reminds me that if I leave the multilanes and travel outside the commercial screen of corrupted land, I can even yet enter a territory where highways and trails open into the country and draw a rambler toward real wandering. Boyer's roads, typically, are strings of macadam or gravel laid narrowly down so that the land can take them in and blend with them. They let us loose to savor the miles of America.

Sometimes I think that our history is about travel and how we've gotten from one place to another and why it was we wanted to get there in the first place. Every citizen in this country traces his oldest forebears to that other hemisphere: we're descendants of wanderers, roamers (blacks might be excepted since they came by coercion), and we all have a peripatetic ancestry that keeps in our blood. Our flag, were it to be a truly graphic symbol, should be yellow stripes on two-lane blacktop, or lined wagon wheels, or a travois on an expanse of green. Did any nation ever build itself more out of transport than have we?

Roadways (think also of rivers, canals, rails) are to Americans as mountains to Tibetans, sand to Bedouins, tundra to Eskimos. Routes have shaped our land, and the land shapes us, so let's pay attention to what molds us. When you look at Winston Swift Boyer's photographs, do you detect kinship and some pride that you live in a land having such places in it? Do you feel enough to act to protect what has been given?

When I go down my twisty and leaf-rimmed lane once made of oak trees, sometimes I wonder whether I like it so much because I can never entirely forget the threats to it and can't ignore its fragility in the face of a swelling population or its vulnerability before real estate Huns whose lone ethic is MAKE BUCKS FAST.

I hope not. I hope I don't cherish something only because it's imperiled, and I hope I'm not a man who can love and act only on behalf of something he's about to lose. But then I look at Winston Swift Boyer's road photographs and find their poignance in that god-awful question, "Glorious, yes, but for how much longer?"

–Columbia, Missouri 15 January 1989

PLATE I *Road to the Sea, Camden, Maine, 1986*

PLATE 3 *Sugarbush Valley Ski Area, Green Mountains, Vermont, 1987*

PLATE 2 *Early Fall, Knox County, Maine, 1987*

PLATE 4 *Northfield Mountains, Washington County, Vermont, 1987*

PLATE 5 *Back Road to South Randolph, Vermont, 1986*

PLATE 6 *Farm Road and White Mountains, Belknap County, New Hampshire, 1986*

PLATE 8 *Road to Old Barn, Belknap County, New Hampshire, 1986*

PLATE 7 *Birch and Mountain Maples, Belknap County, New Hampshire, 1986*

PLATE 9 *4° Fahrenheit, Berkshires, Massachusetts, 1987*

PLATE 10 *Old Yacht Club, Eastern Point Boulevard, Gloucester, Massachusetts, 1982*

PLATE 11 *Eastern Point Boulevard, Gloucester, Massachusetts, 1981*

PLATE 12 *Fort Hill Avenue, Gloucester, Massachusetts, 1987*

PLATE 13 *On Ramp, Highway 95, New Jersey, 1981*

PLATE 14 *Highway 10 with Amish Farmer, Pennsylvania, 1987*

SPEED
LIMIT
45
Simples
Sale
9-4 only

45

PLATE 16 *Blue Mountains, Berks County, Pennsylvania, 1987*

PLATE 15 *Highway 70 with Church, Kentucky, 1987*

PLATE 17 *Farmhouse and Country Road, Preston County, West Virginia, 1987*

PLATE 18 *Highway 80, Columbia, Kentucky, 1987*

PLATE 20 *Fresh and Saltwater Fishing, Darien, Georgia, 1987*

PLATE 19 *Rural Landscape with Tracks, Jackson County, Tennessee, 1987*

PLATE 21 *Oyster Shell Road, Franklin County, Florida, 1987*

PLATE 22 *Plantation Road, The Oaks Plantation, Louisiana, 1987*

Roger Bro

PLATE 24 *"Go West" Marlborough Man and Mississippi River, Quincy, Illinois, and Missouri, 1987*

PLATE 23 *Intracoastal Waterway, Larose, Louisiana, 1987*

PLATE 25 *Dead End, Brown County, Kansas, 1987*

PLATE 26 *Railroad Crossing, Nodaway County, Missouri, 1987*

TONS
15

PLATE 27 *Sunrise, Mason County, Illinois, 1987*

PLATE 28 *Harvest Road, Page County, Iowa, 1987*

PLATE 29 *Frontage Road, Monroe County, Wisconsin, 1987*

PLATE 30 *Lake Michigan, Green Bay, Wisconsin, 1987*

PLATE 31 *Badlands, Pennington County, South Dakota, 1984*

PLATE 32 *Alternate 14, Big Horn County, Wyoming, 1987*

PLATE 33 *Bamforth National Wildlife Refuge, Albany County, Wyoming, 1987*

PLATE 35 *Hans Peak Road, Colorado, 1987*

PLATE 34 *Road to Rawlins, Carbon County, Wyoming, 1987*

PLATE 36 *Ranch Road and Sagebrush, Big Horn County, Montana, 1979*

PLATE 37 *Basque Sheepherder and Bakers Peak, Carbon County, Wyoming, 1987*

PLATE 38 *Battle Mountain, Wyoming and Colorado, 1987*

PLATE 39 *Old Highway 163, Monument Valley, Arizona, 1986*

PLATE 40 *Timpahute Range and School Bus, Highway 375, Nevada, 1986*

PLATE 41 *Yucca Trees, Nevada, 1986*

PLATE 42 *Flaming Gorge Reservoir and Highway 44, Utah, 1987*

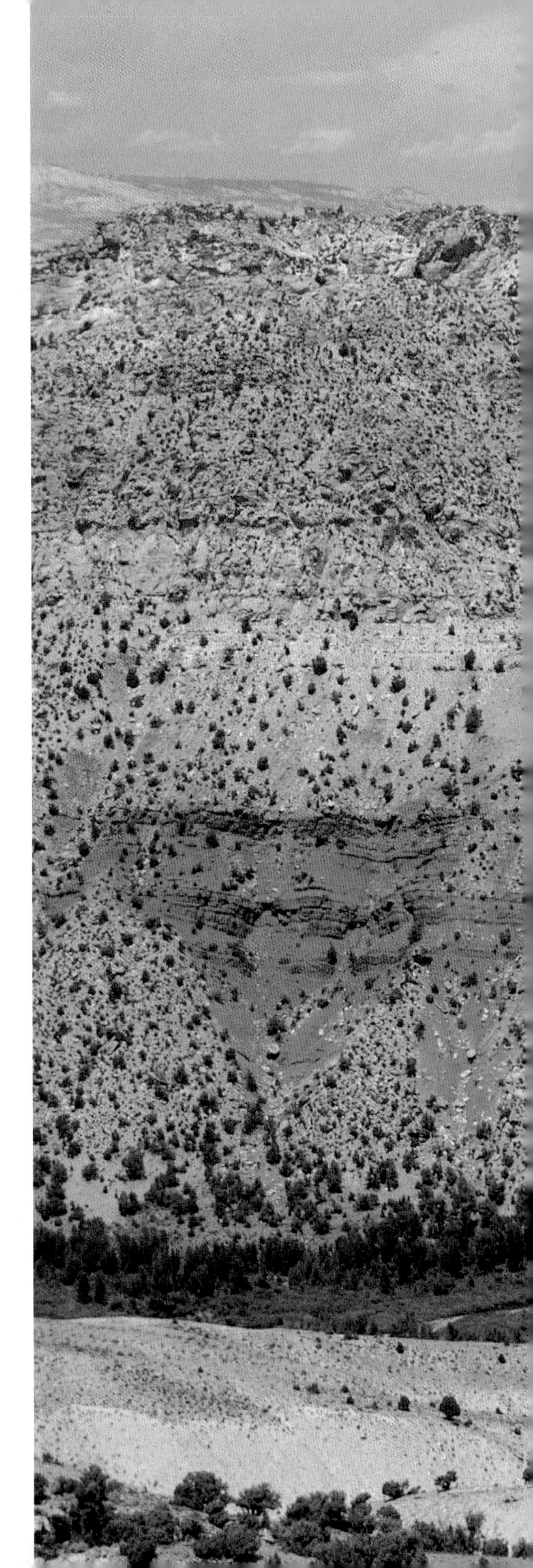

PLATE 43 *Highway 414, Utah, 1987*

PLATE 45 *Courthouse Towers, Arches National Park, Utah, 1986*

PLATE 44 *Highway 9, Zion National Park, Utah, 1987*

PLATE 46 *The White Line, Highway 95, Utah, 1987*

PLATE 47 *Coal-burning Plant and Highway 10, Utah, 1988*

PLATE 48 *Highway 191, Moab, Utah, 1986*

PLATE 49 *Hells Canyon Road, Idaho, 1987*

PLATE 50 *Cape Meares, Oregon, 1987*

PLATE 51 *Causeway, Coos County, Oregon, 1987*

PLATE 52 *Cemetery Road, Lane County, Oregon, 1978*

PLATE 53 *Old Gas Pumps, Lane County, Oregon, 1987*

PLATE 54 *Reedsport Bridge, Oregon, 1987*

PLATE 55 *Highway 120, California, 1986*

PLATE 56 *Sierra Forest Fire, Highway 168, Fresno County, California, 1986*

PLATE 57 *Road and Two Trees, Paso Robles, California, 1979*

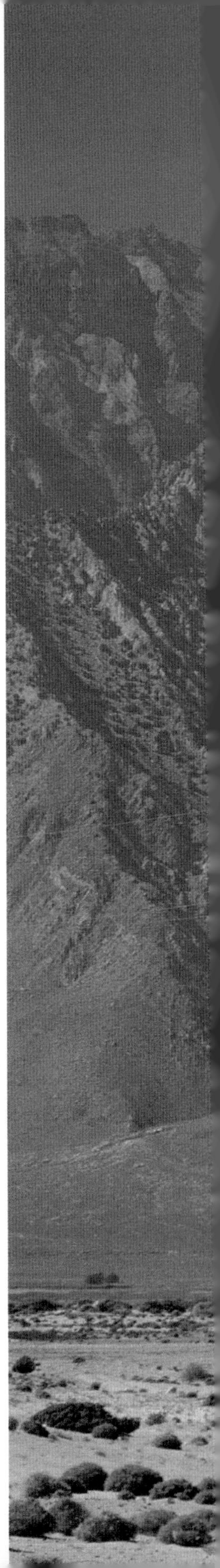

PLATE 58 *Highway 190, Inyo County, California, 1987*

PLATE 59 *Junction 15 and 215, San Bernardino National Forest, California, 1987*

PLATE 60 *Pianalito Canyon, San Benito County, California, 1986*

PLATE 62 *Late Summer, Madera County, California, 1986*

PLATE 61 *Old Highway 101, Monterey County, California, 1979*

PLATE 63 *Highway 1, Monterey, California, 1987*

LIST OF PLATES

PLATE 41 *Yucca Trees, Nevada, 1986*

PLATE 42 *Flaming Gorge Reservoir and Highway 44, Utah, 1987*

PLATE 43 *Highway 414, Utah, 1987*

PLATE 44 *Highway 9, Zion National Park, Utah, 1987*

PLATE 45 *Courthouse Towers, Arches National Park, Utah, 1986*

PLATE 46 *The White Line, Highway 95, Utah, 1987*

PLATE 47 *Coal-burning Plant and Highway 10, Utah, 1988*

PLATE 48 *Highway 191, Moab, Utah, 1986*

PLATE 49 *Hells Canyon Road, Idaho, 1987*

PLATE 50 *Cape Meares, Oregon, 1987*

PLATE 51 *Causeway, Coos County, Oregon, 1987*

PLATE 52 *Cemetery Road, Lane County, Oregon, 1978*

PLATE 53 *Old Gas Pumps, Lane County, Oregon, 1987*

PLATE 54 *Reedsport Bridge, Oregon, 1987*

PLATE 55 *Highway 120, California, 1986*

PLATE 56 *Sierra Forest Fire, Highway 168, Fresno County, California, 1986*

PLATE 57 *Road and Two Trees, Paso Robles, California, 1979*

PLATE 58 *Highway 190, Inyo County, California, 1987*

PLATE 59 *Junction 15 and 215, San Bernardino National Forest, California, 1987*

PLATE 60 *Pianalito Canyon, San Benito County, California, 1986*

PLATE 61 *Old Highway 101, Monterey County, California, 1979*

PLATE 62 *Late Summer, Madera County, California, 1986*

PLATE 63 *Highway 1, Monterey, California, 1987*

DESIGNED BY SUSAN MARSH

PRODUCTION COORDINATED BY AMANDA WICKS FREYMANN

COMPOSITION IN MONOTYPE CENTAUR BY MICHAEL AND WINIFRED BIXLER

PRINTED AND BOUND BY DAI NIPPON PRINTING COMPANY, LTD.